On Gravitation and Electric Charge

Greg Feild

October 29, 2016

The program of physics:

    The program of physics is to devise concepts and laws that can help us understand the universe. Physical laws are constructions of the human mind, subject to all the limitations of human understanding. They are not necessarily fixed, immutable, or good for all time, and Nature is not compelled to obey them.

          -- Richard T. Weidner / Robert L. Sells

In for a penny, in for a pound.

          -- English proverb

Abstract:

In this final paper on quantum gravity, we amend our previous conclusion (see "Observations on the Quantum Mechanical Nature of Gravity") that the motion of mass alone gives rise to the magnetic force. Instead, we conclude that magnetic fields are generated by the motion of both the electric and gravitational charges.

In addition, we determine that the photon and the graviton are the same particle; i.e. there is one (massless) gauge boson that couples to the 'total' charge of a particle; be it mass, or mass plus electric charge.

In this third and final paper on quantum gravity, we will look at a few odds and ends from "A Quantum Mechanical Theory of Gravitational Interactions" and "Observations on the Quantum Mechanical Nature of Gravity", and present a couple of final ideas for the refinements to the standard model suggested in these two previous papers.

Fortunately, not much damage has been done to the current standard model. Most of the formalism, mathematical calculations, and procedures should 'port' quite easily to this simpler formulation.

Basically, we have added a spin 1 graviton and an energy dependent gravitational charge (i.e. particle mass). However, we have replaced the weak charge and color charges with the gravitational charge, and QCD with gravity. Also, this theory allows for the self coupling of virtual gauge bosons if this is still found to be necessary.

In our modified QCD, the linear force term arises from the mass of a particle and is multiplicative; rather than including an additional force term which is linear in the separation R. Also, there is a natural attenuation of this force as particles approach relativistic velocities; hence, it can never become infinite. (As an added bonus, there are no more fractionally charged quarks, summing over all colors, etc.)

In this refined standard model, we have two fundamental fermions; the electron and the electron neutrino; and four gauge bosons; the photon, the W, the Z, and the graviton.

In addition, from time honored arguments of beauty and symmetry, we suggested that the neutrino must (might, could) have a magnetic moment given by

$$mu\_v = e*hbar/2*m\_v*c \qquad (1)$$

The neutrino is now considered a magnetic point charge just as the electron is an electric point charge. Since the neutrino does not carry electric charge, we concluded that the magnetic moment of the neutrino must arise from a 'combination' of its mass and spin.

We then concluded that all (current) induced magnetic fields are generated by the motion of mass and not by the motion of electric charge.

This relationship between mass and magnetism, as well as equation (1) for the muon magnetic moment, were not derived, but were rather two rash and compulsive conclusions. Unification, or bust!

A simple look at the equation for the classical Lorentz force on a particle of charge q moving under the influence of the electromagnetic field of an electron source charge

$$F = q*E + q*(vxB) \qquad (2)$$

shows we cannot simply replace a particle's charge q, with it's mass, m, in the magnetic term. The electrical and magnetic constants, epsilon_0 and mu_0, are buried inside and if we replace the charge with the mass, then the product of these two constants (with mu_0 properly amended) would no longer yield the speed of light.

Instead, we propose, that what we have up to now called e, the fundamental electric charge of the electron, is actually the product of the mass of the electron and it's 'primal' (for lack of a better, as yet untaken, adjective) charge, e'. Here, e' is just the charge to mass ratio of the electron.

$$e = (m\_e)*e' \qquad (3)$$

$$e' = e/(m\_e\_rest) \qquad (4)$$

Where m_e_rest is the rest mass and m_e is the total relativistic mass of the electron.

We can then write equation (2) as;

$$F = (mq')*E + (mq')*(vxB) \qquad (5)$$

We can now add the gravitational force terms due to the source electron mass

$$F = (mq')*E + (mq')*(vxB) + m*F\_g + m*B\_g \qquad (6)$$

where F_g is the usual gravitational force due to the source electron's mass

$$F\_g = G*m1/R^2 \qquad (7)$$

and B_g is the 'gravitational magnetic force', which we are totally not making up, and is given by

$$B\_g = (G/c^2)*m1*(v1xR)/R^3 \qquad (8)$$

where m1 and v1 are the mass and velocity of the source electron.

Obviously, this new force B_g was introduced for symmetry's sake and is based on the corresponding electromagnetic term. The constant giving the strength of the force is chosen to give our new resulting gravitational waves the speed of light.

Now, we want to find the (classical) force per unit mass; or more specifically, the force per unit particle. So, we divide equation (6) by the mass, m, and get

$$F/m = (q/m)*E + (q/m)*(vxB) + F\_g + vxB\_g \qquad (9)$$

Here, the charge to mass ratio is now considered the more fundamental property of a particle. If our particle of mass m, is an electron it will feel all four force terms. If it is a neutrino, it will only experience the two gravity terms. Either way, at relativistic velocities the magnetic forces become appreciable.

This symmetrization of the classical Lorentz force is compatible with our new theory of the graviton and the quantum mechanical nature of the gravitational force.

The Lorentz transformation:

If electric and magnetic fields are due to the motion of the 'product' (and sum) of the influences of mass and electric charge, this would explain why the Lorentz transformations, which were devised to make the equations of electrodynamics gauge invariant, also apply equally well to velocity transformations in particle mechanics.

Electric charge is an invariant quantity and it seems odd (in hindsight) that motion should affect its strength in an interaction.

As far as we know, one cannot have charge without mass even though the Maxwell equations are formulated in terms of 'massless' charge sources. One equation in particular,

$$Del.B = 0 \qquad (10)$$

has been seen as 'asymmetrical' to the other three, and has been interpreted to mean there are no lone magnetic poles. Since magnetic forces now seem to be due solely to the motion of charge, we can interpret equation (10) as a classical statement on the conservation of mass.

Fine structure constant:

Now that the motion of the combined electron mass and charge is responsible for the generation of the electromagnetic force, we can understand the origin of the fine structure constant and why it is not actually a constant.

From the discussion above, we see a proper treatment of electrodynamics would include the contribution of the gravitational charge of the particle as well as its electric charge. Hence, the use the fine structure constant as the coupling constant in QED calculations, rather than just e^2. This is to account for the fact we have neglected the mass term in the coupling. This neglect only becomes apparent at relativistic energies; hence, the suspected running and/or non-constancy of the fine structure constant reported by many researchers.

Gravitons and gravitational waves:

We can now construct a picture of classical gravitational waves.

An accelerated mass will emit waves of varying and alternating B_g and F_g fields, just as an accelerated electric charge emits varying B and E fields.

We note now, of course, that an accelerated electron will emit electromagnetic and gravitational waves; i.e. photons and gravitons. This means all rays of light also contain gravitational waves.

OR, the photon and the graviton are the same particle.

First, we note that we have basically constructed a mass dependent electric charge in equations (2) and (3). Second, both the photon and graviton are massless, spin 1 bosons. They have no extra quantities (or qualities!) such that one should couple to the mass and the other to the electric charge. For example, why does the photon couple to electrons as well as neutrinos? One particle has mass plus electric charge, the other has only mass.

We must conclude there is one massless gauge boson and it couples to the total charge of the particle and does not distinguish between gravitational charge or electric charge but cares only for the product and/or sum of the charges.

This is the main conclusion of this third and final paper on the quantum mechanical theory of gravitational interactions and grand unification!

Now for a few odds and ends.

The edge of the universe:

Previously, we argued if space is three dimensional and our universe began with the big bang, then the universe should have an absolute center and we should be able to determine our current distance from this center.

This means our universe also has an edge. The current radius of the universe would be given by

R_universe = (age of universe)*c

where c, of course, is the speed of light. Once we have incorporated gravity into the standard model and subsequently generated a better estimate for the amount of mass in the universe, we should be able to predict its maximum radius.

What lies beyond the edge of the universe? Undifferentiated space. In other words, the same space we know and enjoy right here, except there is no thing to define it. No objects, no radiation, and just possibly no quantum vacuum fluctuations.

We appeal to a modified Mach's principle, in hypothesizing that the production and subsequent annihilation of virtual particle pairs 'from the vacuum' cannot not actually occur in absolute and empty space with no neighboring matter to induce it. A familiar real world analogy would be Compton scattering, where one assumes the scattering center experiences no recoil in the interaction.

This means we are safe from a second universe popping out of the vacuum outside of our own and then swamping us like a celestial tsunami.

This also means there is only one universe; ours. There is no need for extra universes, or extra dimensions for that matter.

Already out of our depth, but with 24 pages to fill, we now turn to black holes.

Black holes:

In our revised standard model, ordinary matter is composed of electrons, positrons and antineutrinos.

When the mass of a body reaches some critical point and the pressure becomes great enough that the force of gravity overwhelms the electrostatic repulsion responsible for the structure of the protons and neutrons, the constituent electrons and positrons will annihilate producing photons. Of course, we will also have the residual electron antineutrinos.

So, we would have a mix of bosons and fermions. It seems the photons would all fall into the same lowest possible frequency energy state, perhaps defined by the diameter of the black hole, and form a Bose-Einstein condensate. The antineutrinos might either escape, or remain gravitationally trapped, thus giving the black hole it's physical extent (and extra mass) due to the Pauli exclusion principle keeping the neutrinos apart.

Time and space:

We still have black holes, but otherwise time and space are pretty boring again. There is no room for time travel, wormholes, multiple universes, or any of the other speculations of the last century. Science fiction writers may be particularly disappointed as they will need to develop a few new tropes!

For physicists, as well as the general public, this is good news. The world turns out to be more or less how it seems, and we can describe it mathematically without having to abandon our intuitive, ordinary understandings of space and time.

Origin of electric charge:

Now, let's take one more look at our two fundamental fermions from the point of view of symmetry. The electron carries spin, mass, a magnetic moment and an electric charge. The neutrino has spin, mass, a magnetic moment, but no electric charge.

There are 2 possible solutions to this conundrum;

1) The neutrino has a tiny electric charge
2) Electric charge is some kind of 'emergent' property of mass

Since the first idea is ruled out by charge conservation, we will choose the second. It also has a certain beauty.

Let us hypothesize that the electron's magnetic moment *and* its electric charge are caused by the motion of spinning mass.

We shall call this idea 'quantum mechanical electromagnetic induction'. Here, opposite directions of spin would generate opposite electric charges (i.e. either electrons or positrons). The generated electromotive force (Joules/Coulomb) of the spinning mass would manifest itself as an electric point charge once one has taken all the proper mathematical limits, etc. (This calculation is left as an exercise for the reader!)

We assume the rest mass and spin of the electron are constant, so this would imply that electromotive force is quantized like all other physical quantities and processes.

Also, the neutrino would have no charge in this model because it is considered a true point particle without 'a middle'.

In "A Quantum Mechanical Theory of Gravitational Interactions", we derived the mass of the electron neutrino in terms of the mass of the electron and the electric charge

$$m\_v = m\_e/e \tag{11}$$

or

$$m\_e = m\_v*e \tag{12}$$

It is now 'obvious' that the electron is just the neutrino with an electric charge; perhaps it is an "excited" neutrino. They differ in mass-energy by a factor of e and in coupling by a factor of e^2.

If we use equation (11) in equation (1), we can obtain an expression for the neutrino magnetic moment that does not explicitly depend on the electric charge

$$mu\_v = m\_e*hbar/2*(m\_v^2)*c \tag{13}$$

Particle families:

We have replaced quarks with leptons; however, there is not a one-to-one correspondence between the two species. The first generation of particle families is now imagined as follows:

proton   = (u,u,d) ⇒ (e+,e-,e+)

neutron = (u,d,d)   ⇒ (e+,e-,v_e^bar)

electron ⇒ electron

electron neutrino ⇒ electron neutrino

In the standard model, the second generation consists of the strange quark, the charm quark, the muon and the muon neutrino.

The 'charm quark' and anti-charm quark can form temporary bound states in high energy collisions ("charmonium" a.k.a the J/Psi meson) which then decay into a muon antimuon pair or an electron positron pair. The strange quark has yet to be seen in such a bound state. Hence, for the second generation, we make the following substitutions;

charm quark   == muon

strange quark == muon antineutrino

Similarly, for the third generation, we have:

bottom quark == tau meson

top quark   == tau antineutrino

We will assume the relationships concerning mass, charge and spin derived for the electron and the electron neutrino hold for the higher generations as well.

Regarding the muon (and tau), the question still remains, "Who ordered that?".

It is tempting to wonder if the muon is an excited electron. However, if we follow the model we have established so far, the muon would be an excited muon neutrino.

It is also tempting to wonder if maybe, just maybe, there is only one fundamental particle; the electron neutrino.

This electron neutrino would be excited massively to become the muon neutrino. If excited electrically; the electron, and so on. How these excitations occur (and why the electron would be stable) is unclear.

Conclusion:

Neither the gravitational charge (m) nor the electrical charge (e) of a particle is a constant of a particle's motion. They are both dependent on the particle's total mass-energy and now translate identically under the Lorentz transformation. This is due to each force component now having an explicit mass term in the classical Lorentz force equation.

All is mass plus motion; and mathematics.

Still, it's only spin.

-- Jon Butterworth

Life and Physics (hosted by The Guardian)

References:

"Foundations of Electromagnetic Theory", Third Edition
John R. Reitz, Frederick J. Milford, Robert W. Christy
Addison - Wesley Publishing Company

Wikipedia

Recommended:

"50 physics ideas you really need to know"
 Joanne Baker

snarXiv

"A Survey of Physical Theory"
Max Planck

"The Physical Principles of the Quantum Theory"
Werner Heisenberg

Notes:

www.ingramcontent.com/pod-product-compliance
Lightning Source LLC
Chambersburg PA
CBHW061454180526
45170CB00004B/1708